RECONNAISSANCE

AND

SCOUTING

A PRACTICAL COURSE OF INSTRUCTION, IN
TWENTY PLAIN LESSONS, FOR OFFICERS, NON-
COMMISSIONED OFFICERS, AND MEN.

BY

CAPTAIN R.S.S. BADEN-POWELL
(LATE ADJUTANT, 13TH HUSSARS),
ASSISTANT MILITARY SECRETARY, MALTA.

WITH PLATES.

SECOND EDITION

LONDON:
WILLIAM CLOWES AND SONS, LIMITED,
13, CHARING CROSS.
1891.
(All rights reserved.)

PREFACE.

Success in modern warfare depends on accurate knowledge of the enemy, and of the country in which the war is carried on.

Scouts are the eyes and ears of an army, and on their intelligence and smartness mainly depends the success of all operations. The brain and strong arm, the General and his troops, are helpless unless the scouts explain where, when, and how to strike or to ward off attack.

These Lessons are the notes, revised and augmented, of courses of instruction conducted by the author; and he publishes them in the confidence they will be of service more especially to non commissioned officers who are instructing men.

The Lessons form the outline, to be filled in by the Instructor, of a complete course of instruction in Scouting and Reconnoitring.

R. S. S. B-P.

GENERAL INSTRUCTIONS TO LECTURER.

THE following notes for Lessons will explain themselves.

For indoor lessons it will be found expedient to divide the time to be occupied by the lesson into three divisions :—

The first, equal to one-quarter of the whole time, to be occupied by questions on the last lesson.

The second, equal to one-half of the whole, to be occupied by the new subject.

The third, equal to one-quarter of the whole, to be occupied by questions on the new subject.

This method fixes the knowledge acquired, and also shows whether those attending are understanding as they go, and also which of them are capable and which not, of eventually becoming efficient reconnoitrers or scouts.

A large black-board divided by painted white lines into large squares will be found of great assistance.

The Lessons are in outline, but Appendix D contains, in alphabetical arrangement, material with which the Lessons may be readily amplified.

CONTENTS.

	PAGE
PREFACE	iii
INSTRUCTIONS TO LECTURER	v

LESSONS

	PAGE
I. RECONNAISSANCE	1
II. SKETCHES AND REPORTS	8
III. PRACTICAL MAP DRAWING	15
IV. ANGLES, DISTANCES, GRADIENTS, HEIGHTS	16
V. PRACTICAL SKETCHING AND REPORTING—OUTDOORS	25
VI. ADVANCED ROAD SKETCHES—INDOORS	27
VII. ,, ,, OUTDOORS	28
VIII. SKETCHING BY CLASS—OUTDOORS	31
IX. ROAD SKETCHING, MOUNTED—OUTDOORS	31
X. EXAMINATION OF WORK DONE IN LESSONS VIII. AND IX.	34
XI. DRAWING AND COLOURING	34
XII. POSITION SKETCHING	37
XIII. ,, ,, OUTDOORS	40
XIV. THE ENEMY'S POSITION	41
XV. ,, ,, OUTDOORS	44
XVI. SCOUTING GENERALLY	44
XVII. RIDING—SQUADRON SCOUTS	47
XVIII. SCOUTING, MOUNTED—OUTDOORS	50
XIX. ADVANCED MAPS	50
XX. IN CONCLUSION—USEFUL PRACTICE	53

[APPENDICES.

APPENDICES.

App. A. Sketch of Position held by Enemy.
" B. Sketch of Position for Defence.
" C. Headings of what to Observe and Report.
" D. Scouts' Dictionary of Terms and Hints.

RECONNAISSANCE AND SCOUTING.

LESSON I.

RECONNAISSANCE.

Definition.—Reconnaissance is the acquisition of knowledge of the country over which military operations are likely to be carried on, and of the numbers, position, and probable intentions of the enemy: and the better the officer commanding the force knows these particulars, the better he will be able to make his dispositions for attack or defence.

Reconnoitrers.—Men personally fit and specially trained for the work will be selected to undertake this duty, either singly or in pairs, and, as a rule, supported by patrols. They are termed scouts or reconnoitrers.

Their Qualifications.—A scout must be a man of intelligence and pluck, and a good horseman, with confidence in himself, that is to say, one who will not lose his head in a sudden emergency, but can trust to himself to get out of all difficulties, and who is full of "dodges" to meet every kind of incident or accident that may occur.

A man of this kind, enjoying good eyesight and power of hearing, is personally fit to be a scout, but he still requires to be specially trained and, above all, instructed as to what sort of information he is required to obtain, and in what recognised form it is to be recorded or sent in.

Information required.—In general, the kind of information required to be obtained by a scout would be the nature and peculiar points of a certain tract of country, the particulars of the course of some road, or the position of the enemy's outposts, main body, &c.

How reported.—The information gained will be handed in for the information of the officer commanding the force, in the form of a written report accompanied by a "sketch" or map of the ground examined.

Map reading.—Every scout must be able to "read" a map, as previous to starting on a reconnoitring expedition the line of country to be examined would be pointed out to him on a map, and he would also have to find his way by the map and not trust to guides, whom it might be impossible to obtain or injudicious to trust.

Sketches.—Having learned to read a map, it is no difficult task to learn to draw a sketch map sufficient for all practical purposes: it is not intended to be a finished affair or a copperplate map, but merely a jotting down of the chief remarkable points and what are called "military features" of the piece of country, with information about them written on the sketch itself or in the report that accompanies it.

A sketch map is a bird's-eye view of the country, such as one would obtain on looking straight down from a balloon: and that of a scout should be on such a reduced scale as to be all on one sheet of paper, in which the "features," that is hills, roads, towns, woods, bridges, &c., &c., are represented by certain signs that are called "conventional" because they are agreed on beforehand to indicate certain things. Thus all the reconnoitrers use the same

RECONNAISSANCE.

signs and the officer in charge easily compares them. "A reconnoitrer can often say more, and that better, in two lines of drawing, than he could in two pages of writing."

Conventional Signs.—[Exhibit and explain a sheet of conventional signs. See Plate I.]

North point.—As a rule, unless otherwise shown, the top part of the sketch is the north of it, the bottom the south, the right the east, and left the west. Where an arrow is drawn on the sketch, the direction of such arrow shows where the north is.

To find where the north is when you are out and have no compass, notice the position of the sun about midday, and stand with your back to it. Your right hand will be towards where the sun rose, that is the east; you will be facing the north, on your left will be the west, and behind you the south : *vice versâ* south of the equator.

A church generally points east and west, with the great window at east, and tower at west end.

At night the "pole star" will always show you where the north is. To find it you must first recognise the well-known constellation the "Great Bear"; and following up the line of the two end stars called the "pointers," you come upon the pole star, not very brilliant, but itself the lower end of a very similar but much smaller constellation—the "Little Bear."

```
            *
                *   *
              *       * Pole Star

    *
        *
          *
            *       *
              *   Pointers.
```

Scale.—The scale of the map is the proportion its size bears to the size of the ground or area

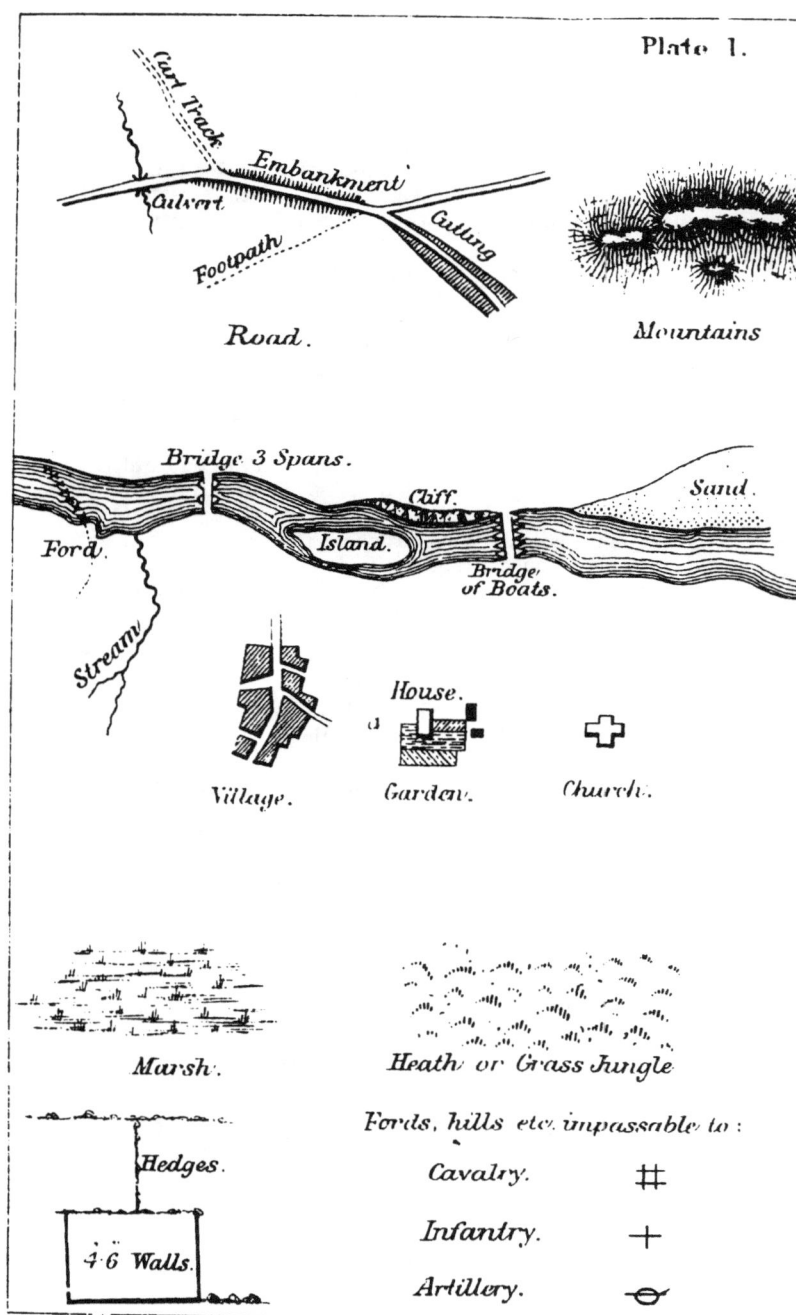

CONVENTIONAL SIGNS.

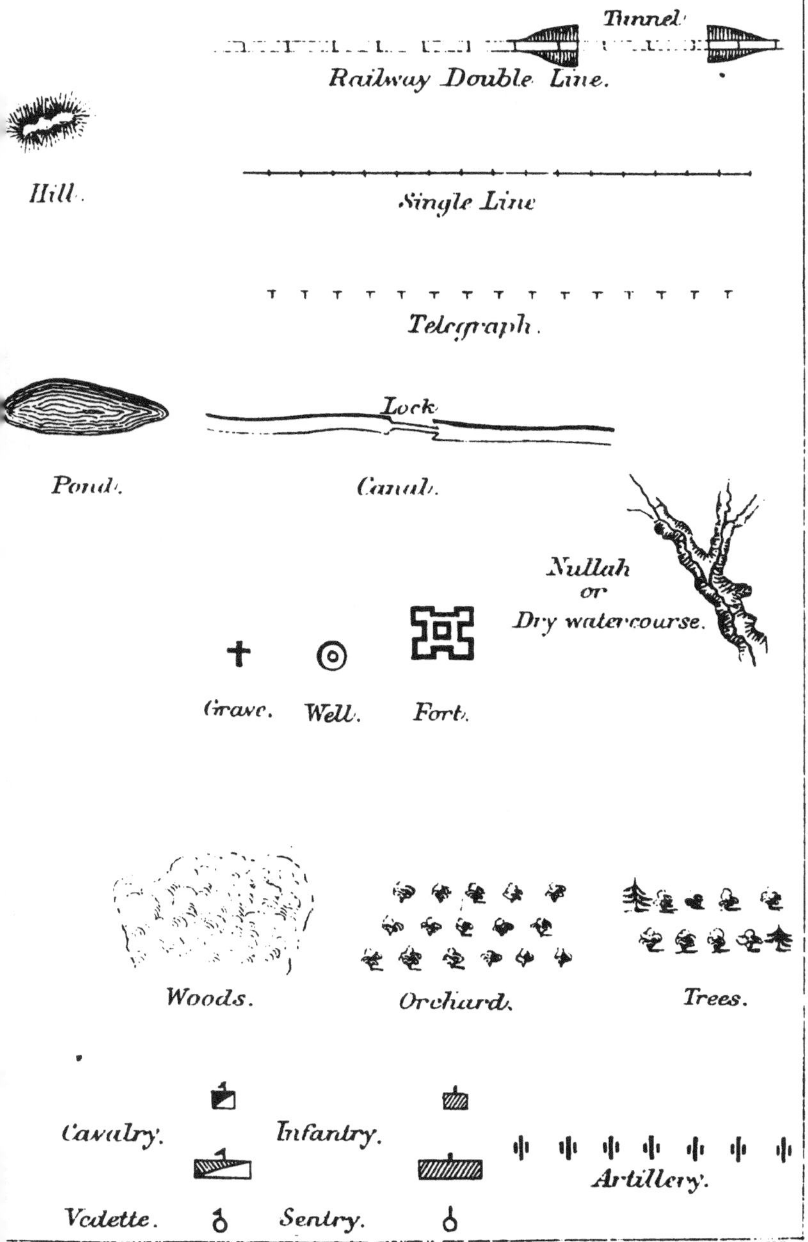
Tunnel
Railway Double Line.
Hill.
Single Line
Telegraph.
Lock
Pond.
Canal.
Nullah
or
Dry watercourse.
Grave.
Well.
Fort.
Woods.
Orchard.
Trees.
Cavalry.
Infantry.
Artillery.
Vedette.
Sentry.

Vincent Brooks Day & Son Lith.

RECONNAISSANCE AND SCOUTING.

represented in it. Thus you will see written on a sketch "Scale, 6 inches = 1 mile"; that means that in six inches of sketch one mile of country is represented. For instance, if a village called Cain is one mile distant from another called Abel, on your map you will find they are six inches apart; and say there is a pond half a mile north of Abel, in your sketch you will find the conventional sign for a pond three inches above the sign for village of Abel.

Military sketches of large tracts of country would be drawn on a scale of 3 or 4 inches to a mile. Sketches of smaller tracts on a scale of 6 inches to a mile. Sketches of villages, positions, &c., on 8 or 12 inches to a mile, showing every small detail, every house, &c.

To make a rough scale for yourself when drawing a sketch, suppose it is to be done on a scale of 4 inches to 1 mile; take a strip of paper or card 4 inches long, which will equal 1 mile, fold it in two, make a mark at the bend, and this mark from end of the paper will give 880 yards ($\frac{1}{2}$ a mile), subdivide this half into half and mark it 440, halve that division and you get 220, and from that 110 yards.

[Draw on the board on a large scale a plan of any spot the men know, such as the barracks or a bit of country in the vicinity, explaining it as you go along, filling in with conventional signs, &c. Question each man on it when finished to see if he can read a map at all. See Plate II.

Give out copies of "conventional signs" for each man to copy, first explaining the more difficult ones and giving hints as to drawing them.]

QUESTIONS ON THE FIRST LESSON.

In a large sketch, which is usually the north of it?

Suppose you have no compass, how do you find the north as you stand, by day? by night?

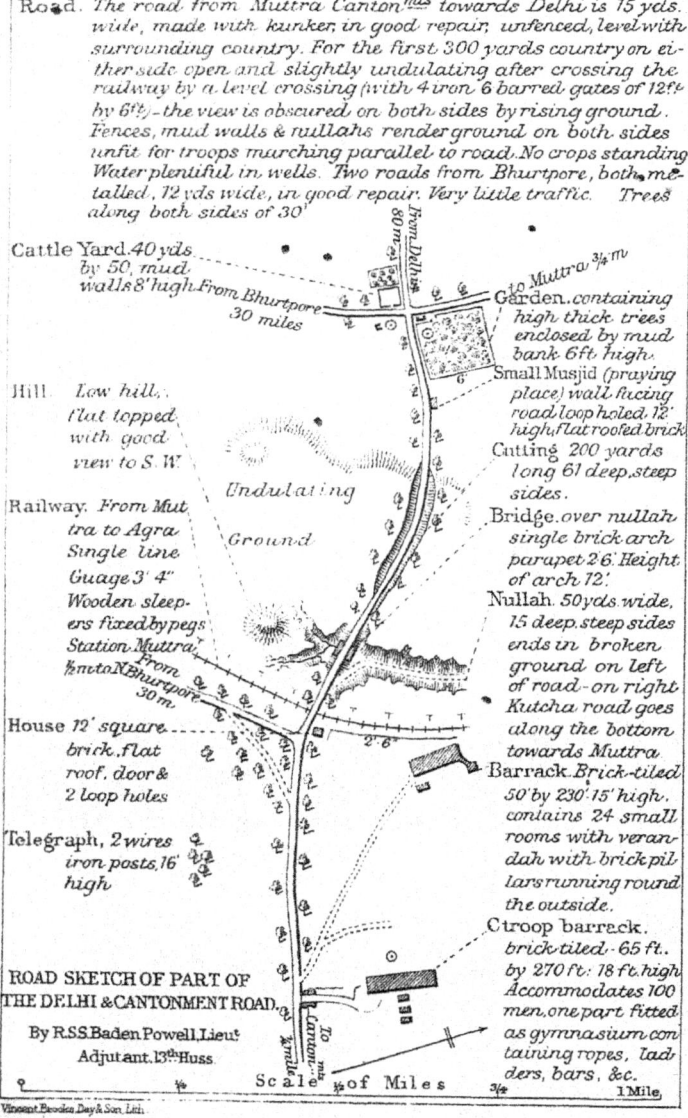

SKETCHES AND REPORTS.

What is meant by a map being drawn on a scale of 6 inches to a mile?
What are "conventional signs"? Which is the right bank of a river? How would you draw a dry watercourse or ravine in a sketch? What is the difference between the conventional sign for a cutting and that for an embankment? What is the difference between the conventional sign for a marsh and that for heath or grass jungle? In the centre of the conventional sign for a river here, an arrow is drawn, what does that mean? Supposing a ford is impassable to infantry and artillery but not to cavalry, what sign would you put against it?

LESSON II.

SKETCHES AND REPORTS.

There are four kinds of sketch and report: 1. The road sketch and report. 2. The field sketch and report. 3. The position sketch and report. 4. The special report.

1. A road sketch is merely a sketch of the road itself, showing the turns it takes and objects that are to be found within half a mile on either side of it. This sketch would be drawn up the centre of the paper, leaving a good margin on each side in which to write descriptions of the various objects on either side of the road as they are passed.

2. A field sketch is a sketch of a tract of country, with the report explaining the various objects and features written out on a separate sheet of paper. In this case the objects would not be described as you meet with them, but would be grouped and described under their headings as in the printed list on the sketching pads. [See Appendix C.] All roads in the sketch would be described one after the other; then all the buildings that were there, and so on with rivers, bridges, and so forth.

3. The position sketch and report deals with all

RECONNAISSANCE AND SCOUTING.

details of locality, forces, &c., of any position held by your own forces or those of the enemy.

4. The special report is one concerned with some one point—as the depth of a ford, the presence or absence of water on a hill, &c.

General Features.—General features include the points that you should notice, supposing you were told to send in a sketch of a bit of country *with* a report. You should state whether the country is flat, undulating, or hilly, if it is open plain, or broken up with watercourses, streams, fences, &c.; what kind of fences prevail, whether walls or hedges, and their height. If the country is cultivated or barren, if the people are numerous and friendly, whether the crops are high and ripe, or not, whether water is plentiful, and if so, how obtained, whether from wells or streams. Description and number of cattle seen, and any signs of the enemy having been in the neighbourhood—e.g. old camping grounds, bits of accoutrements, loopholed walls, &c.

Roads and Paths.—As a rule, a road should be sketched so as to show the principal objects on each side of it for half a mile distance, and particulars of these written up the margin on each side of the sketch. The width of the road should be stated; whether it is macadamised or not; whether in good or bad repair; whether materials are handy for repairing it with. Whether the road is open to the surrounding country or closed by hedges, &c., and whether the country on either side would admit of troops marching alongside the road. Whether it crosses any bridges, hills, passes through defiles, cuttings, embankments, giving their length and depth, and what roads branch from this one, what towns are passed, and so forth. All hills should be noticed and their gradients given. Ordinary carriages cannot ascend anything more than 1 in 7; guns may be got up short hills of 1 in 4.

SKETCHES AND REPORTS.

Towns and Villages.—The names of towns and villages should be carefully found out, and correctly entered in the report. In India a town often has two names (e. g. Quetta is also by some tribes called Shailkote), in such case enter both names: if there is any peculiarity in the appearance from a distance of the village or town, state it, as it will often assist in the recognition of the place: state the length of the sides of a compact place or its general length of straggling, also whether it is on a hill, or level with surrounding country, and on or near a road; what the houses are built of and how roofed, the number of population approximately (reckoning it at five people per house), the names of the mayor or principal inhabitants, the amount of forage, house shelter and stabling available, with details of any particularly large buildings, such as churches, barracks, &c.; also state whether the place is walled or not. In Europe, "village" is used for some cottages with a church, "hamlet" for cottages without a church. Record facts as to water supply—depth of wells and of water in them, their number, usual service of water used by inhabitants, &c.

Buildings and Walls.—In regard to buildings, state whether on a hill or not, and what class of building, whether farmhouses, barracks, what their height, whether thatched or tiled, whether built of brick or stone, and particulars of out-houses, &c., their length, and probable amount of accommodation both for men and horses: for men—allow for one man to every yard of front of building for every single row of rooms. Horses occupy 5 feet of length and 12 of width.

Rivers and Canals.—Their breadth (and I will tell you later on how to measure the breadth and force), depth, force and direction of current, and whether tidal or not. The name of the river, the nature of the bottom, whether sandy, stony, or mud; the banks, whether steep or sloping enough to allow carts or guns to descend into the river, also their height, and especially which of them is highest, and so commands the other. Notice in your report whether the river is likely to alter much in size according to the season of the year. Give names, size, and position of any rivers or streams running into the one you are examining, also description of any islands there may be upon it, whether wooded or not, or inhabited, their size, and height above water; bridges, fords, and locks, and their positions should be given; also the numbers, size (length and width in feet) and description of boats to be found at any places on the banks. The names of any towns or villages on the river.

Woods and Forests.—Their extent, the average height and species of trees, i. e. whether fir, oak, or beech, &c., whether the trees are close together or not; what kind of undergrowth, whether of bushes or grass; what roads, houses, &c., are to be found.

Mountains and Hills.—Their height, character of their surface, whether rocky or wooded, whether the top is flat or pointed, what roads and paths cross over, and whether they are too steep for artillery or waggons. State in what direction a good view can be obtained from the summit, and for how far.

Bridges.—Their length, breadth, height of roadway above the water, material (whether stone, brick, iron, wood), boats, pontoon, &c., the number and nature of supports, height and thickness of parapet, what villages are near and how situated with regard to the bridge. Also notice if any trees, iron rails, &c., are handy for repairing it, should it be broken down.

Railways.—Their gauge (i. e. the exact distance from the inside of one rail to the inside of another), whether a single or double line, length and height of embankments, tunnels, cuttings, bridges, &c. How many telegraph wires, and the height and material

SKETCHES AND REPORTS.

of the posts. The sleepers on which the rails rest, whether wooden beams or iron pans. The names of any stations on the line, and distances to them.

Railway Stations.—Description as under head of "Buildings," length, height, breadth, and material of the platforms; the number of sidings, particulars of tanks, pumps, amount of coal in store, number of carriages, trucks, vans, engines, available. What forges there are, and what stock of spare rails, sleepers, &c.

Ferries.—The size, description, and number of boats available and how worked, whether on a rope and pulley, rowed or punted, &c.

Fords.—Their depth? (Over 4 feet 4 inches is considered impassable for cavalry, 3 feet for infantry, 2 feet 6 inches for guns.) Notice whether the ford goes straight across, or in a diagonal direction; also whether its bottom is muddy, stony, or sandy. (Paths leading down one bank and up the opposite one show where the ford is to be found.)

Marshes.—Their extent, what paths across them there may be, and whether they are to be found at all times of the year, as many bogs are only temporary during wet weather, or irrigation season.

Lakes.—Their extent, depth, description of islands and shores. What boats are available. Are fish plentiful?

Enemy.—In reporting on the enemy, state whether he is stationary or not. The strength and position, arm and dress of his piquets and outposts. If the enemy is posted on a ridge, in a plain or wood, or if entrenched. If the outposts appear vigilant, and sentries plentiful, and patrols numerous. Give, if possible, the direction they take, and their hours of going out. If the approaches to the position are open, and if commanded by guns, or held

RECONNAISSANCE AND SCOUTING.

strongly. Any peculiarity of the ground in front of his position.

All reports should have recorded on them the name of observer, date (including hour), place, and state of weather; and also clear reference to those present when it is made (in case of death or capture of recorder further explanations are needed).

In reports, never give ideas—only facts. Do not say "a wide road," but "a road 15 yards wide."

Always note any distinctive features—as the character of church towers; peculiar colouring; cliffs, trees, &c., of remarkable shape—so as to secure correct recognition by others.

QUESTIONS ON THE SECOND LESSON
(to be asked in the Fourth Lesson).

Suppose you were told to draw a sketch 6 inches to a mile and you had no scale, how would you make a rough one to work with?

What signs stand for feet and inches?

Why is it necessary to show the direction of the north on your sketch?

Why is it necessary to state the scale on which it is drawn?

In first starting to draw a road sketch, where do you commence working on your paper? And how do you begin?

What do beginners often forget to do when they have marked their starting-point and the direction of the road? [To put down the north and south line, and so fix the relative position of the sketch with the actual north.]

After pacing along the original line of the road for say 100 yards, it turns rather to the right: how do you find out the amount of turn to draw it in your sketch? [By first adjusting the sketch with the actual north—getting the north and south line correctly under that of the compass—and then place the edge of the ruler on your present position on the sketch, and move the end of it till it is pointing down the road in the new direction.]

What is meant by "undulating" country?

Plate III.

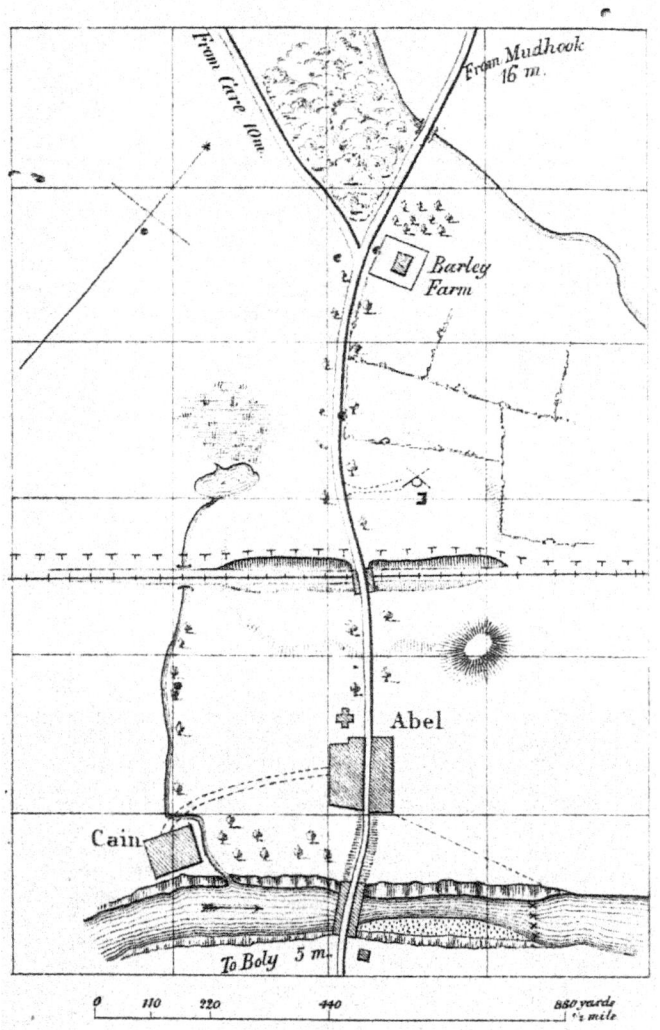

LESSON III.

PRACTICAL MAP DRAWING.

Draw skeleton map of a road on the black-board prepared with squares. Explain how to copy or reduce by using similar and proportional squares. Let the men copy the skeleton map on to their own paper ruled with squares. Let them make paper scales of 6 inches to a mile (as described in first lesson). Dictate features and objects at different distances along, or at right angles off the road, which the men draw in with correct conventional signs without referring to list. Point out that roads should be drawn narrow, writing and trees all one way up, name, date, scale, and north point to be entered; every stroke or dot to have a meaning. Roads entering from top and left to be labelled "From so and so, so many miles"—roads going out to right and bottom "To so and so." Collect these sketches, and keep them for next lesson. See Plate III.

QUESTIONS ON THE THIRD LESSON.

If a map 12 inches to a mile is given you to copy on a reduced scale, say of 6 inches to a mile, how would you set about doing it?

How do you make out a road sketch and report? and how an ordinary sketch and report of a tract of country?

What do you call this latter kind of sketch?

RECONNAISSANCE AND SCOUTING.

LESSON IV.

ANGLES, DISTANCES, GRADIENTS, HEIGHTS.

[Hand out the men's dictated sketches of previous lesson corrected in ink, and pass round your own sketch of the same piece for inspection and comparison. Show specimens of road sketches and reports.]

Angles.—In making out a road sketch, I have only told you to enter each object off the road in the sketch as you come abreast of it on the road; but it

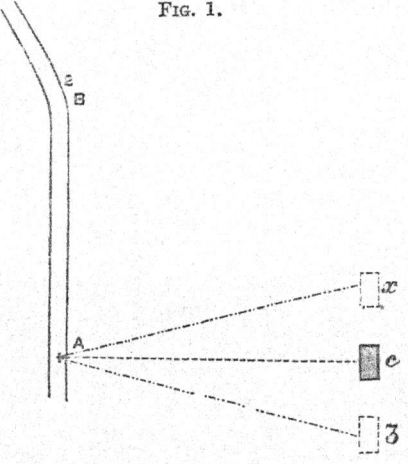

Fig. 1.

is not always easy to tell when you are exactly abreast of an object, especially if it is rather far away. Remember that a right angle is a square angle like the corner of this book; and that for an object to be abreast of you it must be at a right

ANGLES, DISTANCES, GRADIENTS, HEIGHTS.

angle with the line you are proceeding along at the point where you happen to be.

Thus: A B (Fig. 1) is the line you are proceeding along and you are at A. *c* is a house 300 yards away to your right. If *c* is at a right angle with A B, you are opposite to *c*, and would therefore enter it in your sketch. A rough way to find whether you are at a right angle with *c* at A is: Suppose you are facing B; turn half right, close the left eye and look as far as you can to the left with the right, move your head round till the object you were marching on (B) is the last thing you can see to the left. Then, without moving the head, close the right eye and open the left. The last object you can now see to the right will be at right angles to A B: so if you can just see *c* with your left eye, you are opposite to it; if, however, you find you quite easily see *c*, and some ground to the right of it, it shows you are not yet opposite to it (as A *x*), and must therefore move further up the road A B; or if you cannot see it at all, it shows you have gone too far and passed it (as A *z*).

Another and a more accurate way of laying down a right angle, and one which should be employed whenever you have time for it, is carried out by means of a cord (or headrope, reins, &c.) and stakes (pegs, sword, &c.).

Supposing you are standing at P (Fig. 2) and you want to move off at a right angle to the object A: first plant a peg at P, then another between P and A at O, a few paces from P—mark the line from P to O. Then take any point a couple of paces from this line and mark it with the peg C. Then fasten one end of your cord to C and take as much cord as will reach from C to P, walk round with this amount, marking the ground with the end in your hand, thus describing a circle round C. Where this circle cuts the line O P mark it as D, and move round with your circle till you get D and C in line, mark that point E. Join E P, and D P E is a right angle.

RECONNAISSANCE AND SCOUTING.

To be able to find a right angle is specially useful in determining the breadth of a river, or the distance from you of any object you could not pace up to on account of some obstacle.

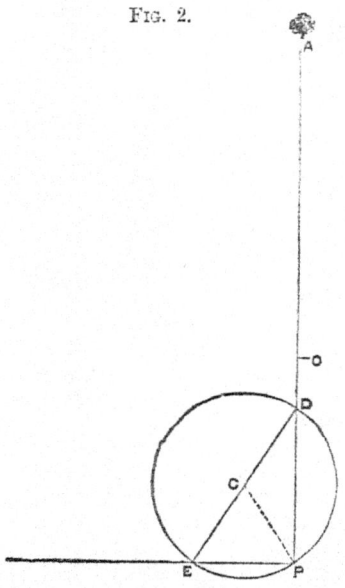

Fig. 2.

Suppose you are standing on the bank of a river at A (Fig. 3) and want to know the distance across: select some object x on the opposite bank; then find an object on your own side of the river at a right angle with x A. Then pace towards it for say 75 yards to B; at two-thirds of the distance, i.e. at 50 yards, a', plant a stick in the ground. At B find some object to your left at right angles to the line you are proceeding along, and pace from B towards it, looking back continually to your stick at a' until you find it comes in a line with x; then the distance

ANGLES, DISTANCES, GRADIENTS, HEIGHTS.

you have paced from B is half the distance of A x. So if you double the distance from B to c it gives you the width of the river at A. This method of

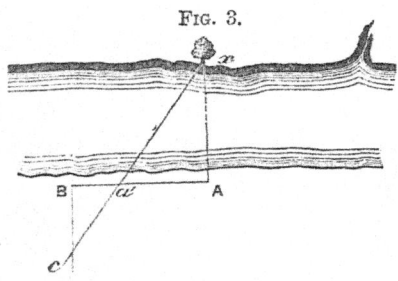

FIG. 3.

finding a distance would also be of great use when reconnoitring an enemy in position, in finding how far off he is from certain points.

To measure velocity of current, pace 60 yards on bank, and time something floating low in the water —so as not to be affected by wind. If this takes 60 seconds passing the 60 yards, the velocity of the current will be at the rate of 2 miles per hour, and so on in proportion.

Gradients.—In describing hills, and particularly when a road goes up or down a hill, you will be expected to give the "gradient" or steepness of the slope. This can either be described in degrees, to find which an instrument is necessary; or, what is more simple, by stating the number of feet in which the hill rises 5 feet. To find this, face the hill, and hold any flat thing, such as your sketch or your hand, horizontally (see Appendix D) in front of your eyes: notice the point where it cuts the ground in front of you: pace up to this point, which will give you the distance in which the hill rises a height of 5 feet; this distance should also be given in feet. Suppose, for instance, you hold up your sketch and pace up

RECONNAISSANCE AND SCOUTING.

to the point where it appeared to cut the ground, and you find the distance is 5 yards, that is 15 feet, you would describe that as "gradient 5 in 15."

The following are the conventional signs used:—

Any gradient under 5 in 4 is   +
"   "   "   5 in 8 is   ⊐⊏
"   "   "   5 in 18 is   -⊖-

So that against the road in the example you would enter -⊖-

Hills.—In sketching a hill, as you go up it take the gradient from the bottom till you reach the top; i. e. when you have finished as far as the spot where your sketch held horizontally in front of your eye first appeared to cut the ground, hold it up again in the same way and move up to the next point, and so on till you arrive at the top; each time that you take the gradient gives 5 feet of height to the hill: add them together, and on arrival at the top you know the height as well as the general degree of slope of the side. Notice the general shape of the top of the hill, whether it is circular, or long and narrow, or what, and draw the outline of it in your sketch in pencil, and from this line draw the shading out on all the sides, with the lines close together where the hill is steep, and wide apart where the slope is easy.

The position of the hill in the sketch should be fixed by taking the bearing of it from two points along your base, the same as any other object.

Hill Sketching.—In proceeding to sketch a hill, first take the bearing of the highest point of it from two points on your base, and so fix on your paper the position of the top. Then ascend the hill, taking the gradient of the slope as you do so. On reaching the top, should you find it to be flat, a plateau as it is called, pace its dimensions, and enter it in your

ANGLES, DISTANCES, GRADIENTS, HEIGHTS.

sketch; if it goes off into a ridge, take the bearing of the direction in which it runs and the distance, noting its breadth at the top.

Take the direction of all streams and watercourses in which they run downwards, and draw them in your sketch, as they are the best guides to the shape of the hill, and show where the heights and where the depressions exist, the watercourses themselves being of course in low ground, the ground between them being higher.

Sketch in the general outline of the bottom of the hill. You should take the bearings of all important objects in the surrounding country, visible from the hill-top; then, in descending, select a different line to that by which you came up, and one which appears to be on the most easy slope, and take gradients as you go. Enter the line of descent and its slope in the sketch.

Drawing and Shading Hills.—For shading in a hill in the fair copy of your sketch, the simplest method is to scrape some black-lead off a pencil, then with a crayon stump rubber, or some chamois leather or even blotting-paper doubled over the point of a pencil, take up a little of the lead and rub it into the sketch along the course of all streams and watercourses, &c., and gradually work from them all over the sides of the hill, but letting the shading become lighter and lighter as you proceed. All sloping ground should be shaded, but only steep places and depressions should be dark, and the rising ground left very lightly shaded. The flat top of the hill and the flat country at the bottom should not be shaded at all. The shading should become lighter near the bottom of the hill, so as to blend gradually with the white of the flat ground.

Cliffs and steep or broken places should be darkly shaded, and then depicted by sharp touches of pen or pencil.

In order to prevent its rubbing off, black-lead

RECONNAISSANCE AND SCOUTING.

shading should be lightly painted over with a thin wash of gum and water, or milk.

In the accompanying sketch of a hill, Plate IV., the line A B represents the base line you are proceeding along. The hill has two points to it, so from y you take the bearings of both points, then pace on to z where you take the second bearings of each and fix them in your sketch. Then from z you ascend the nearest point (dotted line gives your track), noting the gradient as you go. On arriving at the flat top you sketch in its outline, take the direction of various watercourses and sketch them in, and sketch in the general outline of the foot of the hill as far as you can see. Take the bearing of neighbouring hills, or other features, and then proceed to the second point of the hill where you repeat the same operations, after which descend by a new route (dotted line), noting the gradients as you go.

Contours.—The lines in this sketch are called "contours." It is not necessary for you to learn in this course how to make out contours in a sketch, but it is as well that you should know the meaning of them, as they appear in all finished military maps.

The line[*] outside of all the others in the sketch shows the foot of the hill where it joins the flat ground: the first line within this shows where the side of the hill has attained a height of 25 feet: the next line within that shows a height of 50 feet and so on: each contour, therefore, shows a rise of 25 feet in the height of the side of the hill, and any two objects that are on the same contour are at the same height above the ground. Imagine that a flood occur over the country in this sketch and that the water rose to a depth of 25 feet: the

[*] The foot of the hill need not be shown in this way in ordinary sketches; it is only done here in order to assist explanation.

Plate IV.

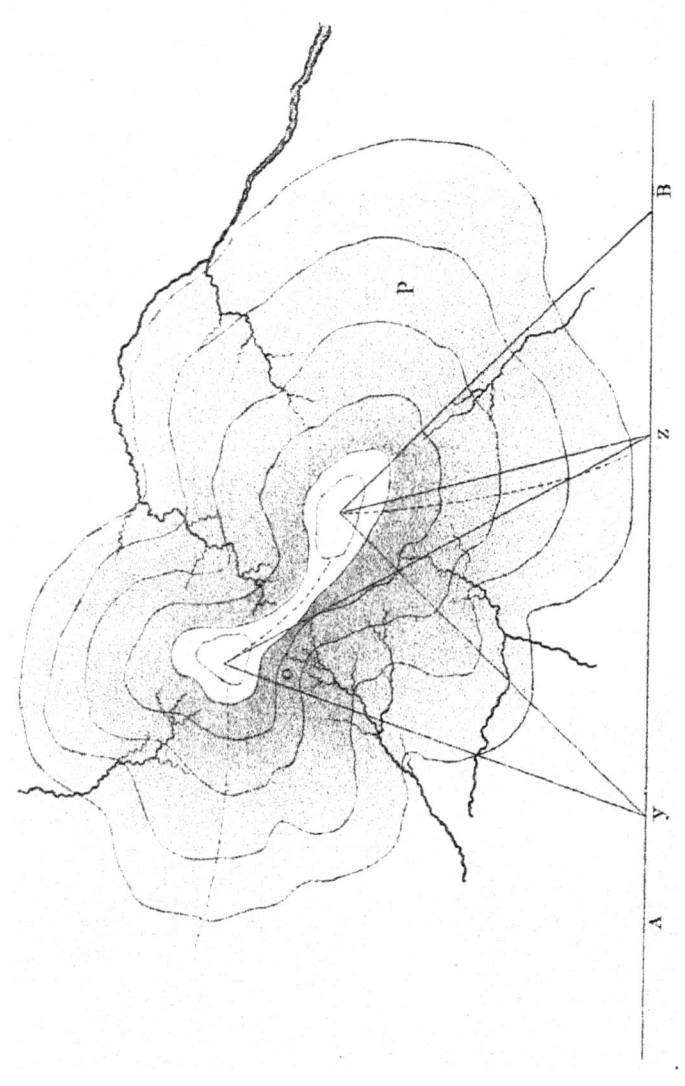

ANGLES, DISTANCES, GRADIENTS, HEIGHTS.

line of the top of the water round the hill would give the line of the first contour, and then if the flood rose to 50 feet of water, the water mark left would show the second contour line.

Thus when looking at a hill in a sketch you can readily ascertain its height in feet by adding together the number of contours on its side and multiplying by 25.

Moreover, you can at once recognise where the side of the hill is steep or otherwise, as, where the contours come close together (as at O in sketch), the slope is steep, and where they lie far apart it is on an easy gradient.

Contours, as a rule, and in maps on a scale of 6 inches to a mile, show a difference in level of 25 feet, but occasionally they are used to show a lesser or a greater height, e. g. 10 feet rise, or 50 feet, but in such cases a note on the map or numbers will tell you of the fact.

To find the height of a building, &c.

Plant a pole b B (Fig. 4) in the ground, upright, at a certain distance from the foot of the object a A; then

FIG. 4.

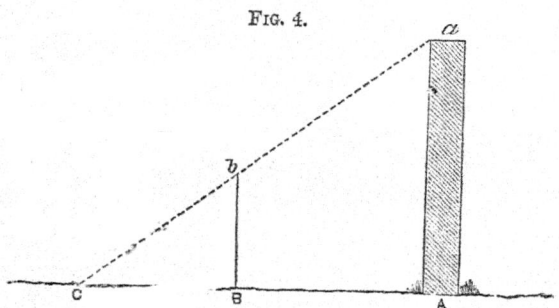

find the point C where with your eye to the ground you find that the top of the pole b comes in line with a the top of the wall. Pace the distance C B, and measure the pole b B, then the proportion that C B bears to *b* B is the proportion that C A bears to *a* A. Measure C A to find height of *a* A. E. g. supposing *b* B were 10 feet high and B C proved 15 feet, that shows *b* B as two-thirds of B C, therefore *a* A will be two-thirds of A C. You measure A C and find it to be 45 feet, then *a* A the height of the wall is two-thirds of 45; that is, 30 feet. When the object to be measured throws a shadow, measure its shadow and also the shadow of a pole, &c., of known length. The proportion this shadow bears to the pole is the proportion the shadow bears to the building, &c. In this, as in other matters, much may be done by practising measurements with one's own limbs. Thus the fingers of the hand held out at right angles to arm, cover at 100 feet distance about 12 feet in height, or a gradient of 1 in 8, and so on in proportion.

LESSON V.

PRACTICAL SKETCHING AND REPORTING—OUTDOORS.

[Take the class out, providing each man with a sketching table, magnetic compass, cardboard map-scale, pencil, sheet of foolscap, and copy of printed "headings for reports" (Appendix C), and let them do a short road sketch.]

At commencing, fold the paper lengthwise in three to define a centre space for sketch, with margin on each side for report on the various objects as they are met with, see Plate V. Make a dot to represent the place where you now stand at the bottom, and up the centre of the paper draw a line pointing from this dot straight up the paper; move the paper round till this line points up the road you are going to sketch; place your compass on a part of the paper that is not likely to be wanted; put a dot against

Plate V.

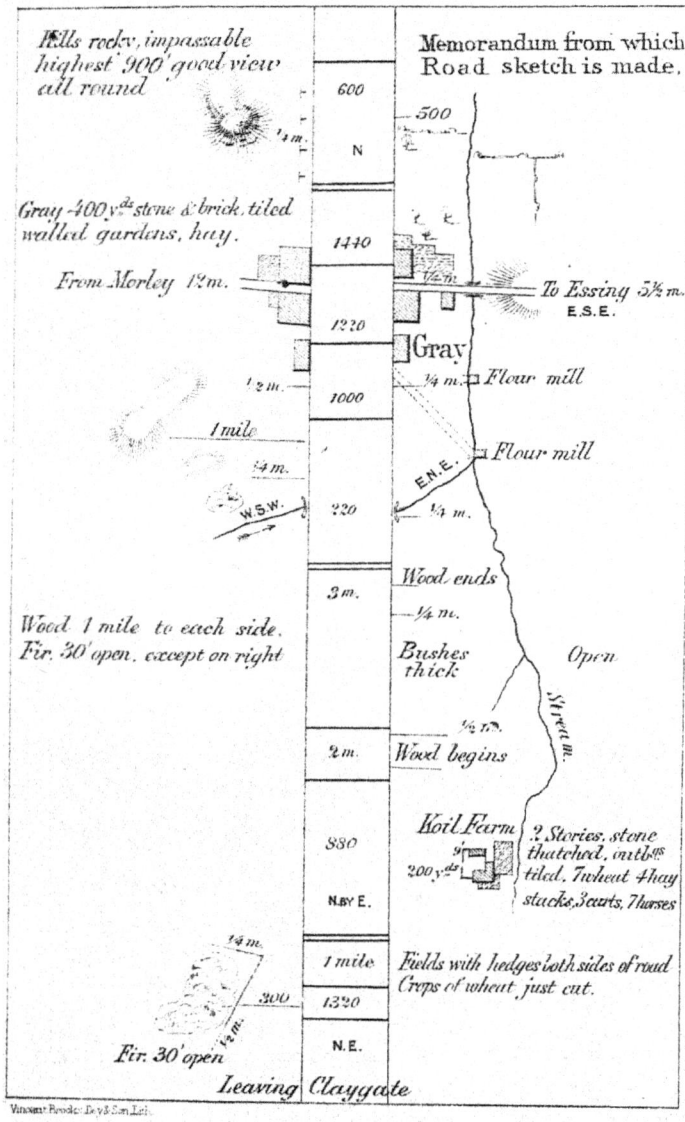

ADVANCED ROAD SKETCHES—INDOORS.

the north and south points, join and produce them; then whenever afterwards the road turns, you will first adjust the paper with the north, placing the N. and S. line under those points in the compass, and then aim down the new direction of road with the dark line on the back of the scale, from the point you have arrived at in the sketch, and draw in that line as the new direction of the road. Objects on either side of the road (to a distance of 300 yards will be enough for this sketch) to be entered in the sketch as you come abreast of them, and described at once in the margin.

LESSON VI.

ADVANCED ROAD SKETCHES—INDOORS.

[Correct each man's road sketch of the previous day, aloud, for the benefit of the rest of the class as well.]

You all seem now to understand how to draw a road sketch, but generally, and particularly on service, the value of a road sketch would greatly depend on the quickness with which it is done. Working as you did yesterday, it would take a long time to put 5 miles of sketch on to paper, yet a road sketch would seldom represent less than that, and would have to be done in about two or three hours and handed in. Well, in the first place, to do this you would be mounted, your horse would do all your pacing and at a faster rate than you could yourself do it; secondly, in an ordinary road sketch on a scale of 3 inches to a mile you would not have to put in half the small details that we put in yesterday in our 8 inches to a mile sketch; and thirdly, the way used for noting down your information would be far simpler. You merely take

RECONNAISSANCE AND SCOUTING.

your sketching pad in your sabretache, and a pencil and compass. You only use this paper with the double line down the centre. That double line represents, in a way, the road you follow; at first starting you notice what direction the road runs in by your compass, and write it between the lines at the bottom of the page. You then make your horse pace along the road while you look out to either side for objects to report on; as you come abreast of each of these you stop your horse, write between the two lines the distance he has come, and then on that side of the lines that the object lies describe it, giving the distance it is from the road as well. When the road changes direction, you write down at what distance it does so and see with your compass what is the new direction, and when found write that down too. From this rough memorandum you would be easily able on your return to camp to draw an ordinary road sketch and report.

To be able to write down correctly the direction in which a road may run or turn, or in which branch roads may run, you should know the points of the compass. They are set out in Plate VI.

LESSON VII.

ADVANCED ROAD SKETCHES—OUTDOORS.

1. Make each man in class measure height, breadth, gradient, velocity of a *current*, practically. Pace a *base*, taking *angles* of objects from each end, adjusting paper to the compass on both occasions first, and draw the objects in on sketch. Test *pacing* over a measured 200 yards (on a range is simplest). Then mount and test the *horse's paces* at a walk, trot and canter, several times over the same distance, to get the average number of paces in which he does it, the rider to enter result in his note-book.

Plate VI.

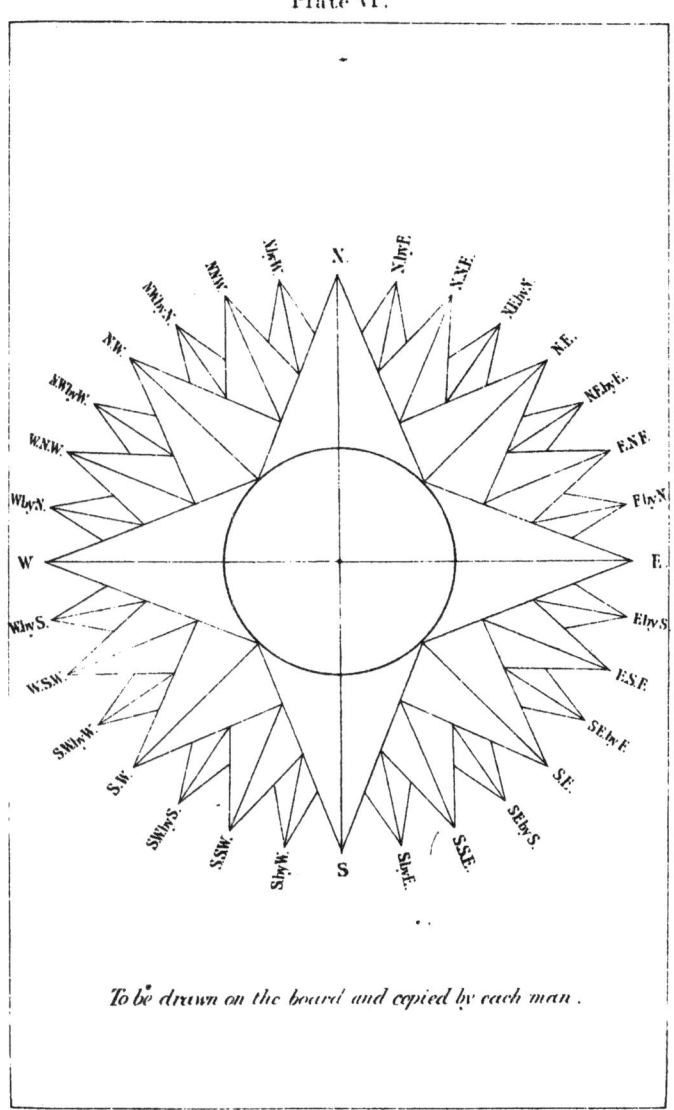

To be drawn on the board and copied by each man.

ADVANCED ROAD SKETCHES—OUTDOORS.

2. Instructor does a bit of quick *road sketching*, using the memorandum, and explains how the information thus collected is to be drawn on paper as an ordinary road sketch, and report on return to camp. The compass would be used in drawing the sketch to give the variations of one point from another, but the needle should be stopped from going round the north by the check, or better still, a circular card made with the compass points accurately marked on it and a hole in the centre. In drawing the sketch, the north point should first be drawn, and then wherever a compass bearing would be required a pencil line parallel to this would be drawn, on which the compass would be laid with N. and S. corresponding with the line. Then, say you want to show direction of a road as N.N.E., put a dot against that point, another against the opposite point, S.S.W.: remove the compass and connect the dots.

3. Instructor draws a field sketch on 8 or 12 inches to a mile, using magnetic compass to adjust the paper at each "station," from which angles of different objects are taken by aiming at them with the ruler laid on the sketch. On the position of an object being fixed by intersection, a man should be sent to examine and bring back all information about it to the instructor, which he enters in his paper of notes. Explains how the lines of intersection would be rubbed out for making the fair copy of the sketch, and the north be made the top of the sketch by making the writing and trees, &c., stand that way: the report would be on a separate sheet, a margin being preserved in which to show the main headings, and the objects being grouped under these headings, numbered or lettered to correspond with numbers written against them in the sketch, and described in the body of the report.

RECONNAISSANCE AND SCOUTING.

LESSON VIII.

SKETCHING BY CLASS—OUTDOORS.

[Outdoors: Field sketch done by the class, 12 inches to 1 mile, using magnetic compass and ruler. Instructor explains how a base should be selected— to lead if possible across the centre of the ground to be sketched; to have a well-defined object at each end, each of which should be visible from the other: the ground between them should be even, to allow of reliable pacing, and should command a view of the prominent objects and features on either side of it: that great care must be taken to move straight along the base: the paper to be adjusted with the north by the compass or with the base by pointing the base line on the paper along the real base, whenever a stoppage is made for the purpose of taking the angle of any object. The first angle to an object being taken before the man comes abreast of it along the base, the second not to be taken till he is abreast of, or has passed it.]

LESSON IX.

ROAD SKETCHING, MOUNTED—OUTDOORS.

[The class parade mounted, taking only sketching pads, compass, and pencils. Make memorandum for a road sketch and report, noting turns of the road by points of the compass and the distance traversed by the horse's paces. To be drawn on return to barracks on a scale of 6 inches to a mile.]

Plate VII.

River Slush. 300 Yds wide, Swift.
   6 to 10 ft deep. No boats
   seen. low sloping banks.

Farley 300 by 200 Yds. Houses stone,
   tiled roofs, on rising ground
   commanding the ford.

Blandford town of 1000 inhab[s].
   1 Mile long, Houses stone.
   tiled Church, Town Hall, Jail.
   Supplies abundant.

Hills rocky, impassable, highest
   900 ft. Road through them
   at Gray very good views
   from hill close to Gray.

Wood extends 1 m. to each side of Road
   and 1 mile along it. Fir trees of
   about 30 ft, Open, except between
   road and stream where undergrowth
   is very thick.

Claygate. village 700 inhabitants on
   junction of Mortown, Tring.
   & Newbridge roads. Stone &
   brick tiled & slated, commanded
   by low hills on both sides.

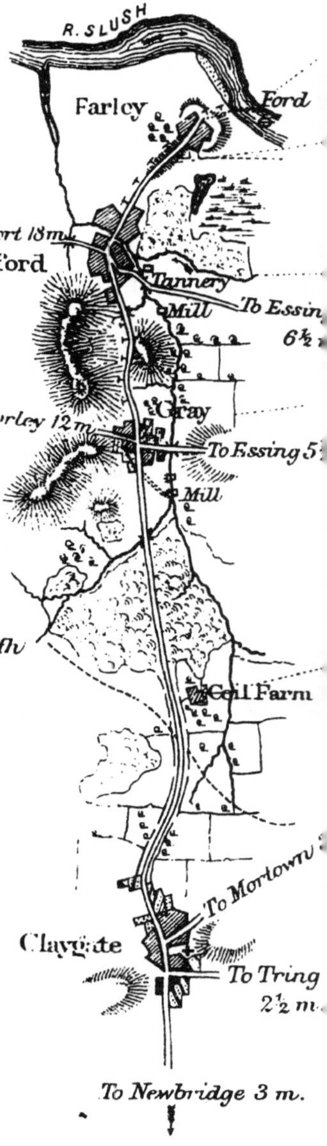

R. SLUSH
Farley — Ford
From Carisfort 18 m.
Blandford — Tannery — Mill — To Essin 6½
From Morley 12 m. — Gray — To Essing 5
Mill
Coll Farm
To Mortown
Claygate — To Tring 2½ m
To Newbridge 3 m.

Scale of Miles

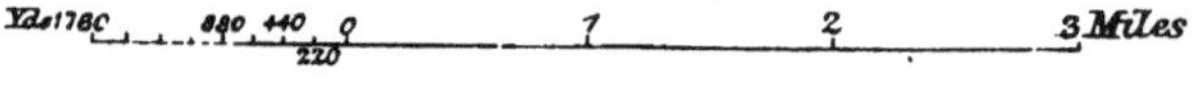

Vincent Brooks, Day & Son Lith.

l gentle slope of bank.
3 f.t deep. Sand & pebbles.

egraph. 3 wires. posts wooden.
f.t destroyed line between
rley & Blandford.

SKETCH OF ROAD
FROM CLAYGATE TO FARLEY.
Reconnoitred by
Lieut. Baden-Powell, Adjutant 13th Huss.

eam. 20 f.t wide 7 f.t deep.
Swift.

ay. village 400 Yds long.
ne and brick. tiled.
led gardens round village.
plentiful, Two flour Mills
the Stream.

General Features

Road is 15 f.t wide macadamized, in
good repair, repairing material ready
all along the road, generally level with
surrounding country enclosed by hedges
for first two miles, where the country
is cultivated, crops of Wheat recently
cut and nearly all carried. Cattle
plentiful in the meadows to right of
road along the course of the stream
Good water plentiful, and easily accessible
in the streams. People fairly friendly
Country on either side of road impassable
for any distance, for all arms. saw
enemy's patrol of 1 N.C.O. 2 privates of
Dragoon Reg.t at Blandford. captured
& brought them in.

m

ble storied stone house.
tched, with courtyard
lled by 9 f.t stone wall.
buildings tiled. 7 Wheat,
hay Stacks, 5 Carts and
orses.

LESSON X.

EXAMINATION OF WORK DONE IN LESSONS VIII. AND IX.

[Correct fair copies of the field sketch and report of the eighth lesson, and the road sketch and report of the ninth; return the last to be redrawn on a reduced scale of 3 in. to a mile, see Plate VII. The instructor to exhibit his own sketches and reports for comparison, and explain that the sketches have been drawn on double the scale that would be usually employed for such sketches, merely because the larger scale is easier for beginners to work on. Road sketches as a rule would be drawn at 3 inches to a mile; field sketches at 6 inches to a mile.]

LESSON XI.

DRAWING AND COLOURING.

Hints for drawing fair copies.—In inking in or drawing the fair copy of a field sketch, it is best, if you have time, and wish to do it neatly, to trace your rough copy made on the ground off on to a fresh, clean, piece of paper. To do this, take a soft pencil and black over the back of the rough sketch under the principal objects shown in it: then pin the rough sketch with its back on the face of the clean piece of paper: go over the lines of the principal features with a hard pointed pencil, which will cause them to be reprinted on the paper below: remove the upper paper and you will find the positions of enough principal objects shown on the new paper to guide you to filling in the rest of the smaller ones. These should all be sketched in

RECONNAISSANCE AND SCOUTING.

lightly in pencil first: and remember that for this, and indeed for all drawing connected with sketches, you must use a sharp-pointed pencil. It is impossible to draw accurately or neatly with a blunt pencil.

As soon as you have shown the exact positions of the various objects lightly in pencil, start to ink them in, for which purpose use a fine nib, but not one that is too fine and needle like.

Where lines are drawn close alongside one another as in the shading of a cutting, embankment, hill, nullah, &c., every stroke ought to be firm and clear: you should never scribble or draw strokes without meaning: you ought to be able to count every pen stroke in the sketch and account for it as meaning something.

In drawing a village it should be carefully shaded

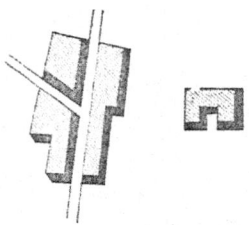

with firm straight lines, all of the same thickness and same distance apart, and lying parallel to each other.

Brick buildings should in the same way be carefully shaded with red. But in drawing a large village it will often be advisable to draw light pencil lines here and there across it, parallel to each other, to serve as a guide for you when shading in in ink to keep your lines parallel then. Moreover, you cannot draw a long straight line without moving your hand, and that is apt to make the line shaky, so it is better to draw short lengths of firm straight

DRAWING AND COLOURING.

lines to form one long one, and, if you like, connect the "joints" of it afterwards.

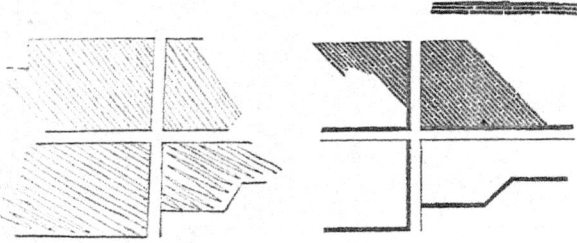

The way it should not be drawn.   The way it should be drawn.

In drawing hills, the steepest parts are shown by lines being drawn closer together. The general outline of the tops of the hills should be sketched in in pencil first, and the shading done in ink commencing from these outlines.

Coloured Sketches.—Should you have to draw a very finished field sketch and colour it, you would show sandy ground with a light coat of yellow, woods with light green, water with blue, brick or stone buildings with red, mud or wooden buildings or villages with black. In a map on a very small scale you may show railways by a red line, and telegraphs with a yellow one. Roads are painted light brown.

First draw in your sketch lightly in pencil; then where you wish to lay on any wash of colour, first go over with a wet brush to damp the surface of the paper, and then take only a little colour in your brush and a fair amount of water, and paint over the damp part of the paper. In this way the colour is laid on evenly, not dark in some parts and light in others. When all the colouring has been done and the sketch is dry you can proceed to draw in the features in ink, or better, with Indian ink. Take care when drawing with ink on any part of the sketch that you have coloured yellow, as the ink is very liable to "run" or blot these.

Any writing or names should be put in small but clear, either in printed or written characters, whichever you find you can do most neatly.

Names of villages should be written close to them conspicuously, so that any one seeing the village on the sketch sees its name at the same time. Names of towns should be written in larger letters than those of villages. Always keep your writing, as much as possible, the same way up as the sketch: you should never have to turn the sketch about to read names, &c., on it.

LESSON XII.

POSITION SKETCHING.

You have now learnt how to do the two important kind of sketches, viz. road sketch and field sketch, with their reports. There still remains to be learnt how to sketch a position.

Sketches of roads or positions are what you would be most commonly called on to draw.

Sketches of positions are of two kinds: 1st, of a position to be held by your own side; 2nd, of a position held by the enemy.

By "position" is meant a spot which from the nature of the ground or other features would be easily defended by a force against another force attacking it.

In the first kind of position, viz. that about to be held by your own people, you would probably receive your orders from the officer commanding your party

POSITION SKETCHING.

as to where to go and what you would be required to find out: e.g. you might be told to go out along the Agra Road till you came to the village Aurungabad, and examine and report on this village to show whether it would make a good position for defending the Agra Road, that is, to prevent an enemy approaching any further along it, from Agra towards Muttra. Before starting on this kind of reconnaissance, always make sure that you understand rightly from which direction the enemy is expected to approach: a "position" should enable the defenders to pour in a fire from both sides of the line of the enemy's advance. On reaching the spot you wish to report on, you should go out to the front for about half a mile and approach it over the same ground an enemy would have to, and notice any houses or hills, &c., in the position from which you can be seen during your approach, and mark them as being the strong points in the position. The sketch may be made either as a road sketch as you advance along the road, or as a field sketch by taking a base line across the front of the position, whichever you find most suitable to circumstances, on scale of 12 or 8 inches to a mile. In either case the report should be made out separate from the sketch.

Remember that information on these three points is most important in the case of a position that is to be defended by your own side.

1st. What are the best points from which a fire could be kept up on an advancing enemy?

2nd. What cover is there for the enemy; especially in the way of sunk roads, ravines, &c., by which he could approach near to, or pass round the flanks of, the position?

3rd. What roads or other lines of retreat are there by which your men may retire should they get driven in by the enemy?

For the first kind of information you must notice the best points as you approach from the front, fix their position in your sketch, and then go and

RECONNAISSANCE AND SCOUTING.

examine each of them in turn, and note down particulars of them.

For the second you must carefully examine the ground for three-quarters of a mile to the front and flanks of the position, and enter half a mile of it in your sketch. But the ground to the flanks should be most carefully gone over, and any hollow roads, walls or hedges, that would conceal an enemy creeping round and past the position should be followed up and reported on. In sketching the ground to the front enter any conspicuous objects in your sketch, such as big trees, solitary houses, rocks, small hills, &c., so that when the enemy has advanced as far as one of these, the officer commanding the defence knows at once from your sketch how far off they are, and can order his men's sighting accordingly.

For the third, find out any roads that lead out of the position to the rear either direct or by a circular route. Also notice any particularly high building or hill that would serve as a good look-out place either for the officer commanding the defence, or for the officer commanding the attack. Also in the case of a village forming the position, show any large building, such as a church, prison, &c., that would serve as a place for the defenders to retire into if the worst came to the worst, and where they could defend themselves till relief came. In the French and German war a great many villages were fought over in this way: the Germans in particular frequently retired into a church when they were driven out of their first position; once in there, they would hold it for hours till more troops came up and relieved them, thus keeping a hold of the village all the time.

In making your report to accompany the sketch, you would state the nature of the position itself, whether consisting of hills, trees, houses, or what. In the case of a village you would go on to describe the village as under that heading. Then give particulars of the ground in front and on the flanks;

POSITION SKETCHING.

then of the roads, &c., for retreat, the chief heights and large buildings; and notice if there are materials handy for barricading the road, such as trees, carts, furniture, beams, &c.

If a pair of you were sent out to sketch a position, the quickest way would be for one to reconnoitre one half of it, and the other the other half, and join your sketches afterwards. In drawing the sketch, the front of the position should be towards the top of the paper. Never mind where the north is, draw the sketch so that the enemy would be approaching from the top of the paper.

Working in pairs.—You would as a rule, out reconnoitring, act in pairs, and you would decide between you, how best to work it according to the different circumstances. In the case of a road sketch of say 5 miles, one would generally do the first 3 miles, and the other would go on to the third milestone and commence his survey there of the last 2 miles: in a position each man would do half: in a field sketch each would work on a different base: but in all cases the fair sketch and report would be made up afterwards by the pair joining their work.

LESSON XIII.

POSITION SKETCHING—OUTDOORS.

[The instructor takes the class out, mounted, gives them their orders to reconnoitre and report on a certain spot as a position, points out the chief features of it, and reiterates the main points of 11th lesson. Tells them off into pairs: each man sketches half the position, on scale of 12 inches to a mile.]

LESSON XIV.

THE ENEMY'S POSITION.

[Indoors. The instructor corrects reduced road sketch and the position sketch.]

I explained to you the other day the requirements of a sketch of position to be held by your own side. This kind of sketch you did yesterday.

Position held by Enemy.—I will now tell you about reporting on a position held by the enemy. In this kind of reconnaissance you will not be able to go close up to his position to sketch it; but this you need not do, the immediate object of your reconnaissance being to find out any cover in the way of hollow roads, ravines, walls, hedges, &c., that will enable your own side to approach his position, or better still, to get round his flanks.

Get as near his position as you can without being under close fire—say 500 yards. Select a base line running parallel to the enemy's line, then take the angles of the different chief points of his position from both ends of your base, or else proceed along the base, noting down the different points as you come opposite to them, and taking their distance from you by the method I gave you for ascertaining the width of a river; but the first would be almost always the best and quickest way. When you have fixed the chief points of the position itself, sketch in particulars of the ground immediately in front of it and on the flanks, and try and find cover leading up to the position—and you must stick at nothing in gaining this sort of information: ride about at a trot as close as you can to the enemy, taking advantage of cover and following it along to see how near the position your own side could approach under it. The enemy may fire at you if they are in a defensive

THE ENEMY'S POSITION.

position, but remember it takes an average of a
man's weight in bullets to kill him, and that is
1400 rounds, there is such missing in action,
especially if the object keeps moving; and if the
enemy are in position as outposts they will not fire
at you, as it would alarm the whole of their line;
they would only send out a patrol to drive you
away, which it should never succeed in doing
altogether; you may ride away from that part of
the line and come back somewhere else a little
farther along. Try and find out particulars of the
uniform and arm of the troops in the position,
and also where they are posted; particularly when
you are reconnoitring outposts in position, try and
see where the piquets are. Remember always to
take advantage of trees, towers, hills, &c., for getting
a bird's-eye view of an enemy's position from whence
you can fill in your sketch if necessary. In this
kind of sketch you should be able to ride over the
ground, noticing the features carefully, carrying
them in your mind so well that on returning to
safety you can jot them all down in your sketch: or
as may very often occur, you may be sent out to go
quickly over the ground immediately before an
attack, examine it, and gallop back and draw a rough
sketch for the officer commanding without having
taken a single measurement or angle. Still, if in
this rapid survey you have found some hidden road
leading round the enemy's flank and can draw a
sketch, however slight, that would give an idea of
its position and course, it would be most valuable to
the officer commanding the attacking force. But
where you have time, a sketch and report should be
prepared, the latter to show, first, the general nature
of the position; secondly, the nature of the ground
in its front and any cover that may be available for
direct or flank movement or attack; thirdly, the
numbers, composition and position of the enemy's
troops; and also in many cases a landscape outline,
giving the appearance of the position and drawn to

RECONNAISSANCE AND SCOUTING.

scale, as it were, with the chief points, their correct distance from each, would prove of great service.

In drawing the sketch put the enemy's position on the upper part of the paper, the ground over which your own people would advance to the attack, being on the lower part.

QUESTIONS ON FOREGOING LESSONS.

What is the usual scale for a road sketch?

Name these points of the compass (pointing to them).

How do you make a road memorandum for a sketch?

In drawing a sketch from the memorandum, say your road begins in a N.N.W. direction, how do you draw it in?

In a field sketch, when do you take the second bearing of an object? Why?

Explain how you would find the breadth of a river without crossing it.

On what scale is a field sketch usually drawn?

How do you find the velocity of a stream?

How do you find the gradient and height of a hill?

Why do you state in reporting on a hill what kind of top it has?

On what scale is a position usually drawn?

What is a position?

What two kinds of position are you likely to have to report on?

With regard to a position to be held by your own side, information is required on three points; what are they?

What are the three points to be found out with regard to an enemy's position?

In drawing a sketch of either kind of position, how do you determine which is to be the top part of the sketch?

LESSON XV.

THE ENEMY'S POSITION—OUTDOORS.

[The class parade mounted proceed to sketch a position supposed to be held by enemy. Starting with a base at 500 yards in front of one flank of position, lay down bearings of chief points; pace across the front parallel to it till opposite the other flank; here take second angles on to the chief objects; fill in detail of ground in front and on the flanks, taking notice of all available cover for attackers. The collecting of information to be done as rapidly as possible.]

LESSON XVI.

SCOUTING GENERALLY.

No soldier on service has such a good chance of distinguishing himself as a scout, and as it depends entirely on himself whether he proves a bold and reliable reconnoitrer, so the rewards and honour for such well-gained information are bestowed on him personally. Therefore in the mean time when once you have learnt what sort of information you will be required to collect and how to report it, as you have done now, you have only to keep up your knowledge of it by practice, noticing the features of any ground you go over, and now and again drawing little sketches of it, especially when off duty.

Nothing should ever escape the eye of a scout; he should have eyes at the back of his head: he should take a pleasure in noticing little trifles or distant objects that have not struck the attention of his comrades. Always notice all peculiar features and landmarks while going over strange ground,

RECONNAISSANCE AND SCOUTING.

especially by frequently looking backwards so that you may be able to find your way back again by them. A scout who loses his way is utterly useless. When scouting on service, keep yourself hidden as much as possible, even from country people, as they will often warn an enemy that you are about. When ascending a hill to get a view, avoid standing up on the top, where you would be very conspicuous, but stand so that you can just see over the top, your body being behind the brow. Keep a sharp lookout for wheel or foot tracks. Also for the glitter of arms in the distance, as you can generally see the glitter of arms in the distance before you can distinguish troops. Flashes slanting mean a force moving across your front in whichever direction they slant down. Upright flashes mean a force coming towards you. Big clouds of dust growing high mean cavalry or artillery. Question country people you meet. Children often tell the truth better than their parents, specially about the presence of an enemy to whom the latter may be friendly.

Remember after passing a difficult place, such as a deep ravine, high fence, thick patch of underwood, &c., to look back at it so that you will be able to recognise the place of passage again at once should you want to come back that way in a hurry, pursued by the enemy, &c.

Make every use of trees and high places for obtaining a view; it gives you a better idea of the country you are in, and with the enemy present you get a good view of his force, and if you are in a tree you are concealed from him at the same time.

As a rule, you should not go more than four miles away from your patrol, and always remember the direction in which your patrol is.

Sketching from Memory.—It is a most useful thing to be able to draw a sketch of a piece of ground, or particularly of a position after having only once ridden over it. It will so frequently occur that

SCOUTING GENERALLY.

you will not have time to go and sketch it accurately, but if you can carry its chief features and their relative positions with regard to each other in your mind sufficiently clearly to put them down on paper in the form of a sketch, such sketch might prove of the greatest value. It is a thing that requires practice to do well, but once learnt will never be forgotten. But the great secret of doing it successfully is to practise yourself continually at noticing peculiarities of the ground and carrying them in your mind.

I recommend you to read a book now in the library, by Mark Twain, called 'Life on the Mississippi.' It is a most amusing book; but the part I wish you to notice is where he describes how he learnt the art of piloting a steamer on the river—the way in which he had to notice little features, and ripples of water, &c., that would escape an ordinary man's eye, and to carry them in his mind. It is just the training that a scout ought to go through.

It is generally in the presence of the enemy and when reconnoitring him that the result of this training proves itself. Suppose you have found the enemy's outposts, but they are too vigilant or well posted to allow you to get within their line for further information: go along their front and notice any streams, hollow roads, fences, &c., along which you could creep at night, and so get into their position. Select several, and notice particular features connected with them by which you will be able to make use of them at night.

Reconnaissance by Night.—When darkness comes on, you may try these ways one after another: see how far into the enemy's ground you can go: find out where his supports and reserves are posted, if his patrols are frequent, regular in their hours of going round, and if they go by certain paths: of what branch of the service the troops are, and their

RECONNAISSANCE AND SCOUTING.

uniforms, pass word, &c. To see an object at night, look just above it rather than straight at it. Remember not to lose your way, but guide yourself by the stars. Try and carry in your head a general idea of the ground, with the positions of the enemy's outposts on it, and the covered paths leading into the position. If you can return to your own lines and make out a reliable sketch of this, you will have succeeded in doing the best piece of work a scout can do.

A night reconnaissance of this kind is best done without your horse.

LESSON XVII.

RIDING—SQUADRON SCOUTS.

"No man is fit for detached duty, particularly scouting, who cannot take his horse cleverly over a fence."

On parade and in the riding-school you are taught to ride your horse on the bit, as from its great power over the horse it enables you to make him keep his dressing, and to turn sharply, &c., as required. In this work the horse, not understanding the words of command, depends on his rider to turn him the right way, &c.; whereas out scouting with rough ground or a fence before him, the horse knows best how to get over it, and in this case the rider to a certain extent trusts to his horse, and therefore ought to give him his head more than on parade to enable him to pick out his way and look where he is going: to do this you should always, when out of the ranks, use the bridoon chiefly, only holding the bit-rein lightly. After a few attempts you will find your horse will go willingly over his fences; whereas he does not like going at a jump straight from parade, because he is afraid of an extra tug on the bit, which is painful to him, while the bridoon is not.

In going at a fence, do not take the reins up too short, lean the body as far back as you can, cling with your thighs, knees, and calves, and draw your feet well back. It is no use leaning the body back unless you draw your feet back too. Use both hands with the reins, and keep them low.

Be careful not to bucket your horse about out scouting; once a man gets into the way of flying about the country, it is very hard to break him of the habit: it does not perhaps do much harm at a field-day, when the horse is sure of a day or two's rest afterwards; but on service, when he may have to do 20 or 30 miles next day, and perhaps again the day after, you would very soon find yourself looking out for a fresh mount.

Go straight up or down a steep place or side of a hill. If you go sideways and your horse slips, it is hard for him to regain his footing; you risk breaking your own leg. If you go straight up, your horse can only stumble, at the worst; going down he may slide, but can generally recover himself. Always dismount up or down very steep places; it saves the horse marvellously.

Swimming Horses.—If your horse in crossing a river suddenly gets out of his depth, let go his head altogether; lean a little forward with your legs drawn back, and hang on by the mane.

In starting to swim a river, it is best to take off your clothes, and let the horse tow you across by his mane or tail; guide him by splashing water at his head. Remember, a horse is easily upset, and a man with his kit on is easily drowned with a little mismanagement.

A horse may be hobbled with a stirrup-leather, which should be arranged for the purpose, so that a slip loop may be made round one foot, and the two ends buckled close round the other foot.

RECONNAISSANCE AND SCOUTING.

The whole of the instruction given you so far refers to the employment of scouts searching for an enemy, or gaining information about him or the country over which operations will be carried on: but they would also have duties to perform as scouts when the enemy has been found and is being encountered or manœuvred against. In this case scouts of each squadron are merely sent out a short distance in front of the regiment more with a view to warning it in time to save it getting unexpectedly into bad ground or coming suddenly upon a concealed body of the enemy, than of obtaining information.

The scouts should be from 200 to 500 yards in advance of the line, and never out of sight of it. When the main body is moving fast, the scouts should be further to the front than when it is moving slowly. They must be on the look out "backwards" for all changes of front of the regiment and conform to them.

On coming to bad ground, scouts will halt, face the regiment, and "recover" swords at any point whither their squadron could pass over. Should there not be such a passage, they should point with the sword to the nearest way of going round.

On coming across the enemy suddenly, scouts would halt and circle on the same principle as vedettes.

Should the regiment be in column when scouts are ordered out, the scouts of the leading squadron extend to the front, those of the remaining squadrons moving out to the flanks: all right troop scouts to the right flank; left troop men to the left.

Scouts trot in on "recall" sounding, and take their places in the ranks. Should the regiment advance to charge without recalling the scouts, they should continue examining the ground to the front till the "charge" sounds, when they clear the front at a gallop and fall in in rear.

In the Franco-Prussian war a regiment of cavalry

advanced to attack some infantry without sending out scouts to the front. Just as they were coming up to the enemy they got into some bad ground, of which they had had no warning, and were cut to pieces by the infantry fire before they could extricate themselves.

LESSON XVIII.

SCOUTING, MOUNTED—OUTDOORS.

[Take the class out, mounted and in patrol "cross" formation, over a line of country in which some easy fences, &c., are to be found: notice and correct defects in their riding. Lead them to the front of a position; suppose it to be held by enemy; show them how to get as near as possible to it, taking advantage of all cover; point out landmarks to guide them in the event of making a reconnaissance under cover of darkness, and show what chief points and features they should take particular notice of, with a view to making a sketch of the position from memory on their return to quarters. Notice trifling details as you go along, and question the men, to see if they have noticed them too.]

LESSON XIX.

ADVANCED MAPS.

[Instructor corrects memory sketches drawn from previous lesson.]

Now so far you have been instructed in drawing sketches in the quickest and roughest way, but in a way best adapted for service. It may at times, however, occur that you will have plenty of time for making a sketch, and that a very accurate one would be required. For this you would require more accurate instruments to help you than an ordinary compass and a ruler, such as the prismatic compass and protractor.

Prismatic Compass.—The actual system of making a sketch with this compass is exactly the same as you have been learning, only you do not have to adjust your paper with the north on taking a bearing; and instead of taking the bearing of an object by aiming at it with your ruler, you aim at it through the two sights of this compass. The hair in the flange being as it were the foresight, when you look at the object through the two sights you will see that a number shows itself below the object; that number tells how many degrees the object is away from north. I take the bearing of that house, for instance: when the compass steadies itself I read 215°. To express this on my sketch: supposing I stand at this dot, I draw a north and south line through it, parallel to that in the corner of my sketch; then I take this ruler, which is called a "protractor"; on it are marked all the degrees of the compass, of which there are 360. That is to say, the circle of the compass is divided into 360 parts called degrees. No. 1 degree is the first to right of the north point, and No. 2 on the right of that, and so on till they come round to the north again, which is No. 360. Well, I place the protractor on my paper with the side of it against my north and south line and on the left of it with this mark on it (opposite 90th degree) against my position in the sketch.

Protractor.—Look along the numbers on the protractor till you come to 215°, and put a dot on your paper to mark it; remove the protractor and join this dot with that showing your station, and it gives you the bearing of the object.

ADVANCED MAPS.

The protractor has two rows of figures on it: well remember that if the bearing of an object is anywhere between 1° and 180°, to mark it on your paper you would lay your protractor on the right of the north and south line and look along the figures nearest the edge for the number you require: if the bearing is over 180° up to 360° you put the protractor on the left of the north and south line, having turned it round so that 360° points to the north, and then read the inside row of figures.

The object of using these instruments is to obtain accuracy, so you cannot be too careful when working with them. Wait till your compass is perfectly steady before noting the number that shows up under the object, keeping the hair sight directly on the object, and read the number carefully to half or even a quarter of a degree, for although halves and quarters are not marked in most compasses or protractors, still you can generally judge the amount by the eye.

Then in "laying off" the angle with your protractor on the paper, be careful that you draw the north and south line through your station exactly in the same direction, parallel with the guiding north and south point in the corner of your sketch and then that the back edge of your protractor lies exactly on this line with the mark opposite 90° just on your station. Use a sharp-pointed pencil to dot the degree or half degree required, and mind that the line you draw to join your station and the degree cuts through both exactly.

In doing a field sketch with a prismatic compass and protractor, your north and south line should point straight up and down the paper.

[Make each of the class take and "lay off" a few angles with aid of these instruments.]

RECONNAISSANCE AND SCOUTING.

QUESTIONS ON FOREGOING LESSONS.

Suppose you are ordered to reconnoitre an enemy's line of outposts as soon as it gets dark, what preparation should you make while there is still daylight?

What is the use of "squadron scouts"?

Suppose as squadron scout you come across a bad piece of ground with one way across it, how do you show this road to the squadron following you?

How many divisions are there in the circle of the compass, and what are they called?

With what instrument do you "lay off" the angles on your paper when you found them by the prismatic compass? How do you lay them off?

Why in reconnoitring a position should you examine the ground for some distance to both flanks of it?

Why in making a sketch of a position to be defended, is it a good thing to enter in your sketch any conspicuous objects in the front of it, such as solitary trees, houses, &c.?

LESSON XX.

IN CONCLUSION—USEFUL PRACTICE.

[Men in pairs using prismatic compass make a field sketch and report of a piece of country, without any assistance from instructors. To give in both fair and rough copies of the work done.]

[Men in pairs without prismatic compasses sketch and report on a position.]

[Men in pairs make a road sketch and report of 5 miles without prismatic compasses.]

[On any occasion when the regiment is practised at outposts at night, take the scouts out to front of the line, and let them try and creep in between the vedettes and gain information as to where piquets are posted, &c., and to make sketches from memory, of the ground and disposition of the troops.]

[Take small parties of men consisting of four or five, for rides about the country, finding the way by Ordnance map, and noticing details and incidents en route.]

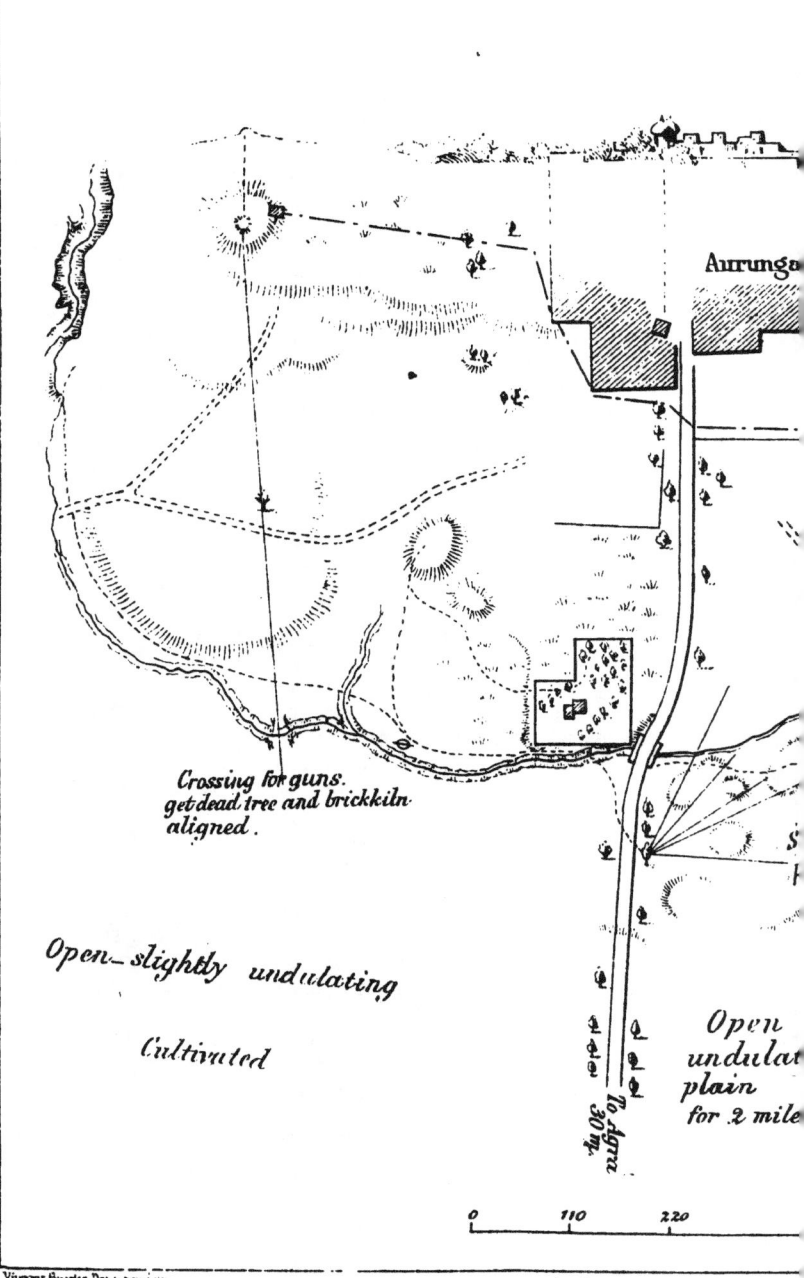

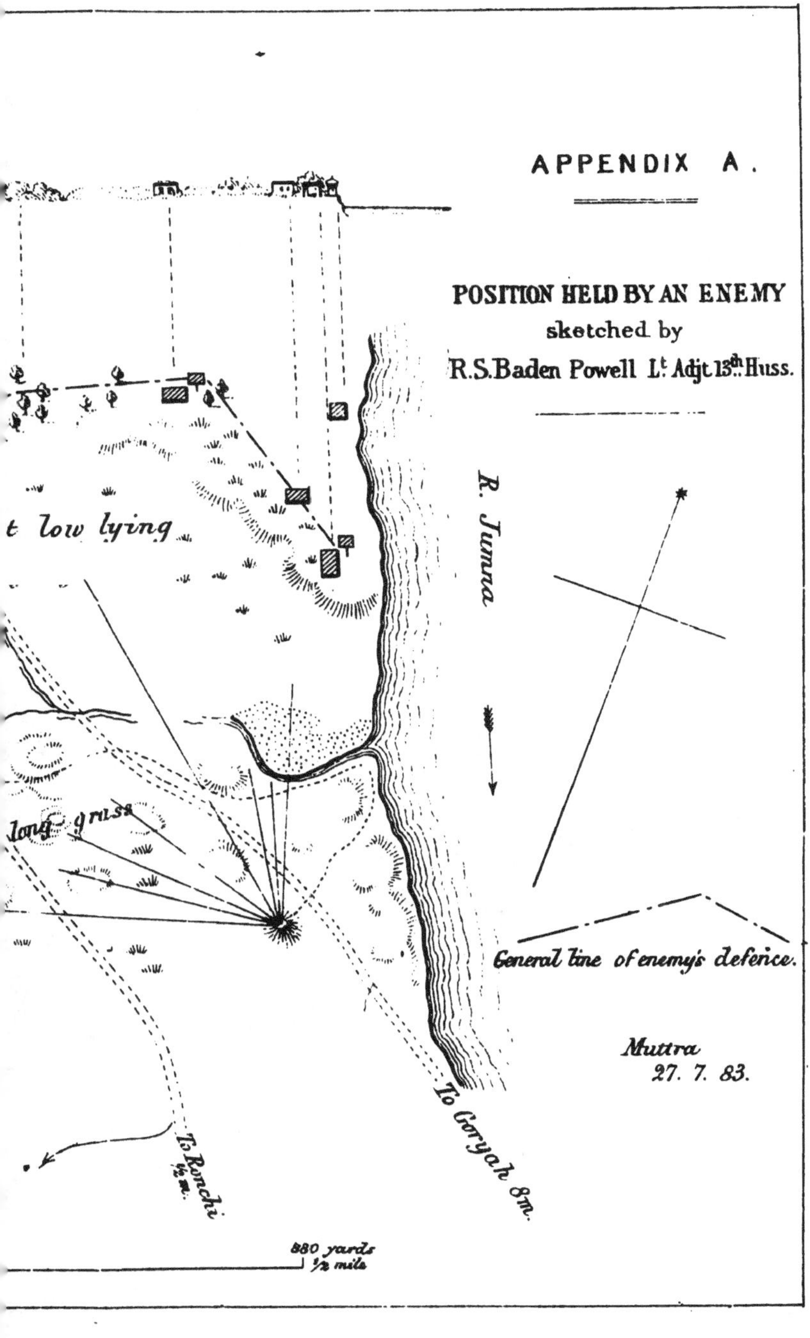

APPENDIX A.

POSITION HELD BY AN ENEMY
sketched by
R.S. Baden Powell Lt. Adjt 13th Huss.

low lying

long grass

R. Jumna

General line of enemy's defence.

Muttra
27. 7. 83.

To Goryah 8m.

To Ranchi ½m.

880 yards
½ mile

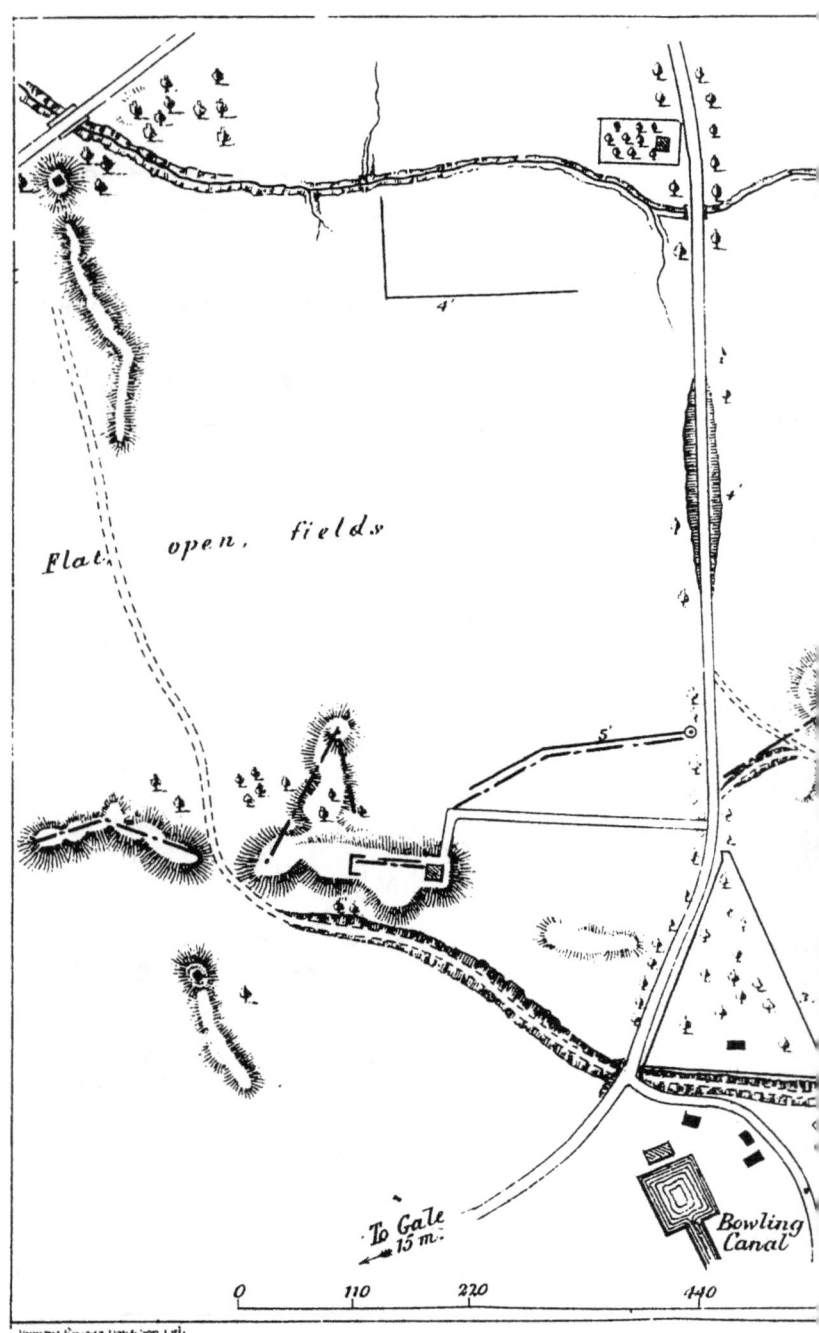

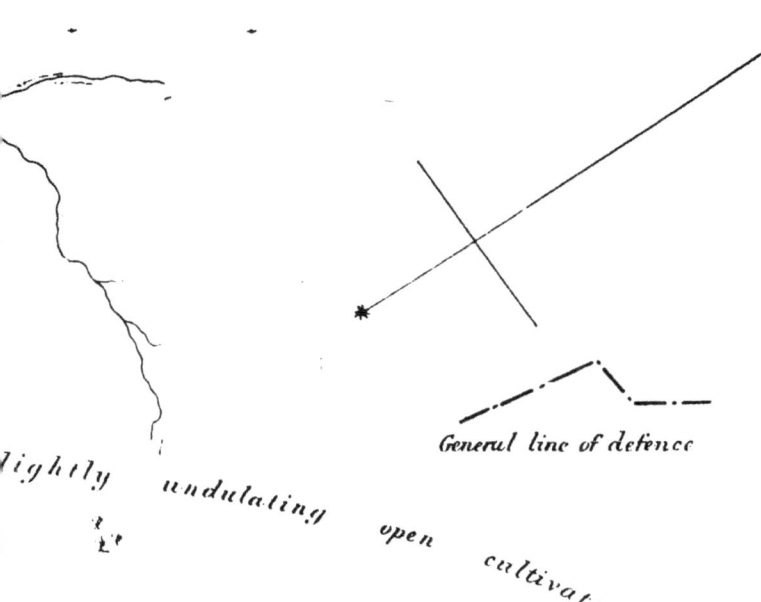

General line of defence

lightly undulating open cultivated

APPENDIX B.

POSITION FOR DEFENCE
OF
GALE AND TAUNTON ROADS AND THE HEAD
OF THE
BOWLING CANAL.

Reconnoitred by R.S. Baden Powell.

L! Adjt. 13th Hussars.

20. 6. 82.

980 yards
½ mile

APPENDIX C.

Headings of what to Observe and Report.

I. Land (for marching across, manœuvring over, or living in).
   (a) *General character of country.* Cultivated or not, crops, kind of fences; number of cattle, cows, sheep, pigs; area and kind of forests, woods, trees, underwood.
   (b) *Obstacles*—watercourses, fencing, heavy soil, or sand.
   (c) *Elevations*—flat or hilly, contours, gradients, ravines, character of surface (whether rocky, grass, &c.), relative heights of hills and knolls, distance and direction of views from top, recognisable features.

II. Water (as being obstacle or aid to advance, or for drinking).
   (a) *Lakes and arms of the sea*—extent, depth, exposure; where fordable; water, whether salt, brackish, or fresh; fish, kind and quantity, nets or weirs; vessels, barges, boats; piers and landing places.
   (b) *Rivers*—breadth; character of bottom and banks; relative heights of banks; speed of current, whether tidal, signs of floods; tributaries; islands; depth and direction of fords and character of approaches; bridges; particulars of boats, ferries, locks.
   (c) *Swamps and ponds*—how far passable; accessible for drinking and watering horses; permanent or temporary.

III. Works (as useful for defence, communication, or bivouac).
   (a) *Roads*—where from and to, width of metal, condition, material, fenced or not, bridges, defiles, cuttings, embankments, levels, gradients, where commanded, cross roads.
   (b) *Bridges*—length; width; height of arch above

APPENDIX.

water; over what; material; supports; means of repair near.
- (c) *Railways*—gauge, number of rails; embankments, cuttings, tunnels; sleepers, bolts.
- (d) *Railway Stations*—material and dimensions of buildings and platforms; sidings; tanks, pumps; stores, coal, spare rails; rolling-stock; forges; general situation.
- (e) *Telegraph*—from and to where; number of wires; height and material of posts; offices, instruments.
- (f) *Buildings*—dimensions and material of walls and roof; doors, windows; out-buildings; stable accommodation.
- (g) *Towns and villages*—name; means of recognition; area; situation; accommodation, number of houses; stores of food; nature of water supply; general material of walls and roofs; number of vehicles, horses, beasts of burden; any factories or workshops.
- (h) *Detached buildings*—similar points; also number of ricks and stacks; head of cattle, &c.; forage.

IV. PEOPLE (as giving information or providing supplies).
Character, whether reliable, friendly, communicative; language; occupation; names of head people; numbers; well off or not; usual foods.

V. ENEMY—all traces of enemy.

APPENDIX D.

SCOUTS' DICTIONARY OF TERMS AND HINTS.

CALCULATIONS: scout has nothing to do with—he collects data only, e.g. reports number of houses, out-buildings, &c., in a village, not estimate of how much accommodation.

Reconnoitrer collects data, and makes his calculations when he has opportunity afterwards.

APPENDIX.

CURRENT. Its velocity is well judged by the following formula:—Pace 50 yards on bank, and time object as it floats past. If it takes—

100 seconds or 130 beats of pulse, current = 1 mile an hour, "sluggish."

25 seconds or 30 beats of pulse, current = 4 miles an hour, "rapid."

17 seconds or 20 beats of pulse, current = 6 miles an hour, "torrent."

COLOUR. In sketch-maps using colour saves much time, especially if in washes with a brush. Always make column on sketch of patches of colours used, and write on each what each means.

DISTANCES. Practise perpetually judging distances by sight: make out various formulæ, as number of beats of pulse between flash and report for every 500 yards of distance; holding hand or finger at arm's length, know how many feet upright it covers at certain distances; mark distances of focusses on field-glass.

ENEMY: signs of. Look for graves and hospitals to show whether there is disease; the presence or absence of heavy guns or trains, and large stores of ammunition or provisions indicate whether he intends advancing or retiring. Arrangements for sleeping, whether tents or not, show intention of remaining or not.

ENCAMPMENT or BIVOUAC Reports. Before starting to make these, obtain calculated space required. In selecting, look for security against surprise or attack; ease and regularity of plan; healthy soil and pitch; absence of obstacles; best available supply of water, fuel, forage and food; facility of access; natural shelter.

FORMULÆ. Scouts should work out formulæ of their own for judging distances, heights, and gradients; speed of currents; pace of selves, horses, &c.; number of infantry, cavalry, &c., that pass a given spot in a minute at a walk, trot, &c.; proportion pulse bears to seconds of a watch.

FORTIFICATION. Sound elementary knowledge of field fortification; of practical methods and means of defending villages or posts, and of entrenching; is necessary in reconnaissance in order to know what to look for and what to report upon.

GUIDES Treat them well; *promise* them liberal pay when work is done; make it quite secure they cannot escape; converse with them freely, asking after their

APPENDIX.

families and occupation; be cautious as to believing; never let them see you know they are deceiving you, till afterwards; if inhabitants refuse to act as guides, carry one off with you. When by himself he will work well for pay.

INHABITANTS. If in a foreign country, have printed lists of questions and make leading inhabitants fill in answers in writing. If they refuse, carry one off with you. If inhabitants are nervous and humble, enemy is far away; if confident and bumptious, enemy is near and in force.

LANDMARKS. Scout should learn to notice landmarks, i.e. peculiar features that others will recognise from description. He should also be able to make them, e.g. by "blazing" trees; marking gate-posts; burning a hedge or tree, and making marks on the ground, noting each on his sketch.

MEASURES. Every good scout knows what part of him or his arms measures 1 inch, 6 inches, 1 foot, 3 feet, or 5 feet. He knows exact height of his eye from ground; length of his pace; width of extended arms.

MEASURES, WEIGHTS and DISTANCES in a foreign country should be recorded in words used by informant. Good scouts have learned the names and equivalent in English, and give the reduction to English terms in brackets.

MEMORY. Never trust it longer than absolutely necessary: always write down every item as soon as possible—if only in the roughest way.

MESSAGES. If *written*, read over before sending; tell purport to messenger that he may destroy document if in danger; write inside hour of sending, and outside by whom sent and whether urgent: if important, send duplicate by another route.

If *verbal*, make messenger repeat it to you before starting.

NIGHT-WORK. Work up time and points of compass by various constellations—which differ in position with time and place: notice what shapes and colours are easiest recognised: work slowly and without noise; things are seen best in the dark from below against the sky line. Avoid animals, as being sure to expose you.

NUMBERS. On the average 250 infantry (in fours) or 120 cavalry (in sections) pass given spot at a walk in a minute; calculate numbers by timing how long force takes to pass.

ORDERING RECONNAISSANCE. In each case the main object should be clearly defined, in order to save time and

APPENDIX.

labour expended on acquiring unnecessary information. Thus if the object be a rapid march right through a district, there is no need for details as to supplies or defensibility of buildings, &c.

PACE. Every good scout finds out what speed his ordinary walk is, and what that of his horse at walk, trot, and hand gallop; timing over mile: also length of stride. He can thus measure distances by pacing or by time.

PRISONERS. Question them on what they know best, e. g. *when* they started rather than where from; mix up many personal questions of interest to them, such as where born, whether Colonel popular, whether they have had enough to eat and drink; what it was; whether their quarters were comfortable. Information as to amount and kind of force they belong to, or of forces they know to be near, will be thus more readily got at and more reliably.

RECONNAISSANCE IN FORCE is an attempt to obtain information about the enemy by pushing in his screen of outposts and vedettes, but is an ordinary military operation for the purpose of enabling those detailed for the duty to make use of the skill and knowledge they have acquired by a course of lessons such as that contained in this book.

REPORTS (*a*) *written*.—Give facts rather than ideas or opinions; if latter, add "think" or "heard from so and so"; write "wall 12 feet high," not "high wall"; not "material, inflammable," but "material, wooden walls and thatched roof"; write legibly; read over carefully before dispatching.

(*b*) *verbal*.—Say over to yourself beforehand what you are going to say; ask to be cross-questioned; speak slowly and with thought.

RIGHT ANGLE.—To find roughly, close eye near to one object, look with other at it, turning head till object is just visible over bridge of nose, say to right: keep head still; close left eye and open right; look over nose to left, and last object visible marks right angle.

RIVER, width of, may be obtained roughly as follows: hold end of stick, rifle, &c., under arm, slanting upwards in front to level of eye; align tip with opposite shore; turn round on heel till tip aligns with some object you can pace to; distance of that object will be width of river or ravine.

SKETCHING. Begin by putting in certain conspicuous features in each section near middle, and give distance;

E

APPENDIX.

next mark in roads, lines of watercourses, lines or water shed; after these put in leading outlines, and contour then colour or shade general characters, as cultivation woods or open ; finish off by details.

SOUND travels 1090 feet in a second—that is roughly ⅓th of a mile. In judging distance by interval between a flash and report, remember that roughly speaking there are eight beats of a watch, four beats of a pulse, or three seconds for every 1000 yards of distance.

On a still day rifle shots may be heard 4000 yards off.
,,   drums   ,,   5000   ,,
,,   artillery firing   ,, 35,000   ,,

STARTING. Before starting obtain in writing the general object of the reconnaissance, and special character of information required; and, above all, limit of time available.

Also obtain all information you can of where you are going, and such rough plan or map as may be obtainable.

TRACKS. Watch for all tracks ; examine all patches of soft ground; be able to tell which way horses or vehicles have gone by their tracks; in coming to soft patches in a road it may often be advisable to make a detour and ride or walk over it in the opposite direction, and then continue by detour again so as to mislead pursuers.

TREES. Scouts should be instructed in chief kinds of trees; their uses as material or fuel; ease of felling; good and bad qualities. The same with regard to the food plants, crops, &c., of the country

www.ingramcontent.com/pod-product-compliance
Lightning Source LLC
Chambersburg PA
CBHW070408230526
45471CB00006B/2698